Sensing and Supporting Communications Capabilities for SPECIAL OPERATIONS FORCES

Committee on Sensing and Communications Capabilities for
Special Operations Forces

Standing Committee on Research, Development, and Acquisition Options for U.S.
Special Operations Command

Division on Engineering and Physical Sciences

NATIONAL RESEARCH COUNCIL
OF THE NATIONAL ACADEMIES

THE NATIONAL ACADEMIES PRESS
Washington, D.C.
www.nap.edu

THE NATIONAL ACADEMIES PRESS 500 Fifth Street, N.W. Washington, DC 20001

NOTICE: The project that is the subject of this report was approved by the Governing Board of the National Research Council, whose members are drawn from the councils of the National Academy of Sciences, the National Academy of Engineering, and the Institute of Medicine. The members of the committee responsible for the report were chosen for their special competences and with regard for appropriate balance.

This is a report of work supported by Contract H92222-07-D-0014, DO #2, between U.S. Special Operations Command and the National Academy of Sciences. Any opinions, findings, conclusions, or recommendations expressed in this publication are those of the author(s) and do not necessarily reflect the view of the organizations or agencies that provided support for the project.

International Standard Book Number 978-0-309-13760-7
International Standard Book Number 0-309-13760-8

Limited copies are available from:

Division on Engineering and Physical Sciences
National Research Council
500 Fifth Street, N.W.
Washington, DC 20001
(202) 334-3111

Additional copies available from:

The National Academies Press
500 Fifth Street, N.W.
Lockbox 285
Washington, DC 20055
(800) 624-6242 or (202) 334-3313
(in the Washington metropolitan area)
http://www.nap.edu

Printed in the United States of America

COMMITTEE ON SENSING AND COMMUNICATIONS CAPABILITIES FOR SPECIAL OPERATIONS FORCES

JOANNE ISHAM, *Chair*, L-1 Identity Solutions
THOMAS BURNS, *Vice Chair,* SET Corporation
KENNETH BOWRA, Oak Ridge National Laboratory
REGAN EDENS, Ares Systems Group
CHIP ELLIOTT, BBN Technologies
CAROLYNE HART, Sandia National Laboratories
ROBERT HENRICK, Arbitron
PAUL HOFF, BAE Systems (retired)
LINDA KATEHI, The University of Illinios at Urbana-Champaign
ROBERT McCLATCHEY, McClatchey Associates
TAMAR PELI, The Charles Stark Draper Laboratory
ED PENTALERI, Loral Communications
W. DAVID SINCOSKIE, University of Delaware
JONATHAN SMITH, University of Pennsylvania
L. DANNY TROMP, The MITRE Corporation
DAVID WHELAN, The Boeing Company
WINSTON WILEY, Independent Consultant

Staff

CARTER W. FORD, Study Director
GREGORY EYRING, Senior Program Officer
KAMARA E. BROWN, Research Associate
URRIKKA B. WOODS, Program Associate

STANDING COMMITTEE ON RESEARCH, DEVELOPMENT, AND ACQUISITION OPTIONS FOR U.S. SPECIAL OPERATIONS COMMAND

JACQUES GANSLER, *Chair*, University of Maryland
A. MICHAEL ANDREWS, L-3 Communications
KENNETH BOWRA, Oak Ridge National Laboratory
LAWRENCE DELANEY, Titan Corporation (retired)
CHIP ELLIOTT, BBN Technologies
MICHAEL HOPMEIER, Unconventional Concepts, Inc.
JOSEPH MANCUSI, Narus, Inc.
RANDALL MURCH, Virginia Polytechnic Institute
WILLIAM NEAL, The MITRE Corporation
ROBERT NOWAK, Defense Advanced Research Projects Agency (retired)
ALTON ROMIG, JR. Sandia National Laboratories
PAUL ROSENSTRACH, The Charles Stark Draper Laboratory
JULIE J.C.H. RYAN, George Washington University
HARRY SCHULTE, Raytheon Missile Systems
ANNETTE SOBEL, University of Missouri
JOHN SOMMERER, Johns Hopkins University Applied Physics Laboratory
DWIGHT STREIT, Northrop Grumman Corporation
DAVID WHELAN, The Boeing Company

Staff

MICHAEL A. CLARKE, Director
CARTER W. FORD, Program Officer
DANIEL E.J. TALMAGE, JR., Program Officer
MARGUERITE SCHNEIDER, Administrative Coordinator
NORMAN M. HALLER, Consultant
KAMARA E. BROWN, Research Associate
URRIKKA B. WOODS, Program Associate

Preface

The U. S. Special Operations Command (SOCOM) is one of nine combatant commands directly responsible to the Secretary of Defense. As a functional command, SOCOM's primary responsibilities are to organize, train, and equip special operations forces (SOF) provided to the geographic combatant commanders and to U.S. ambassadors and their country teams. Currently, SOCOM has a central role in planning and executing the worldwide campaign against terrorism.

Of particular interest to SOCOM are advanced sensors that are able to operate in all environments and that are able to exfiltrate information to overhead platforms and forces in the field. The ability to sense, discriminate forces, locate, and track militarily significant objects has been the goal of numerous science and technology activities since the Vietnam conflict. Emphasis on the need to develop these technologies has not abated and, for SOF to accomplish their mission, this requirement remains important and largely unfulfilled.

SOF missions typically involve the insertion of small, highly trained teams into targeted areas, which may be required to operate independently for days to months. Their success depends heavily on situational awareness and the ability to gather information and transmit it to centers where it can be processed and analyzed. Thus, technologies for sensing and supporting communications (S&SC) are particularly critical. Examples include technologies for remotely monitoring the use of clandestine trails and camps or determining the location and movements of high-value targets and then transmitting this information back to those who can use it to make decisions about tactical responses.

STATEMENT OF TASK AND COMMITTEE APPROACH

Given the rapid advances in S&SC technologies, SOCOM's acquisition executive, in September 2007, discussed with the National Research Council's (NRC's) SOCOM Standing Committee the need for increased capabilities for sensing and communications that could be used by SOF. The NRC agreed to undertake a series of studies to evaluate those that are most promising and assess their readiness and performance in a variety of environments. The original plan was to accomplish a one-year study with emphasis on 1-year of the most challenging environments—triple-canopy jungle. Later, the sponsor requested that additional emphasis be placed on other operating environments that SOF might face. It became evident to the committee that the need for additional in-depth assessments would require more time.

This report offers observations on a limited number of technologies in the jungle environment that the committee has investigated to this point. In this report, the tasks are these:

- Discuss the unique requirements of SOF for sensing and supporting communications in relevant SOCOM operational environments and provide an

introductory-level discussion of sensing and supporting communications issues and challenges associated with effectively operating in these restricted and nonrestricted (open) environments.
- Discuss the sensing and supporting communications architecture and the associated capabilities and technologies required for effectively operating in the jungle/triple-canopy environment.
- Discuss preliminary observations, issues, and challenges associated with sensing and supporting communications in a jungle/triple-canopy environment.
- Identify additional capabilities and/or technology areas that need to be addressed through additional committee data gathering to support any subsequent reports.

The months between the committee's last meeting and the publication of the report were spent preparing the draft manuscript, reviewing and responding to the external peer review comments, editing the report, and conducting the required security/public release review necessary to produce this version of the report, which does not disclose information as described in 5 U.S.C. 552(b). It was mutually determined by SOCOM and the NRC that the full report contained information as described in 5 U.S.C.(b) and therefore could not be released to the public in its entirety.

The committee thanks all who contributed to this effort, including the individual committee members who volunteered considerable time and energy, the supporting staff of the National Academies, and the many outside experts who provided necessary information to inform committee deliberations. Finally, the committee greatly appreciates the opportunities to interact with the knowledgeable SOCOM representatives, who shed much light on how their forces operate and emphasized the importance of this work to their warfighters.

Joanne Isham, *Chair*
Thomas Burns, *Vice Chair*
Committee on Sensing and Communications Capabilities for Special Operations Forces

Acknowledgment of Reviewers

This report has been reviewed in draft form by individuals chosen for their diverse perspectives and technical expertise, in accordance with procedures approved by the National Research Council's Report Review Committee. The purpose of this independent review is to provide candid and critical comments that will assist the institution in making its published report as sound as possible and to ensure that the report meets institutional standards for objectivity, evidence, and responsiveness to the study charge. The review comments and draft manuscript remain confidential to protect the integrity of the deliberative process. We wish to thank the following individuals for their review of this report:

Ruth David, Analytic Services, Inc.,
Neil Fox, Defense Advanced Research Projects Agency,
Larry Lynn, Defense Advanced Research Projects Agency (retired),
C. Paul Robinson, Sandia National Laboratories (emeritus),
Alton Romig, Jr., Sandia National Laboratories,
Paul Rosenstrach, The Charles Stark Draper Laboratory,
Neil Siegel, Northrop Grumman Mission Systems, and
George Sutton, SPARTA, Inc.

Although the reviewers listed above have provided many constructive comments and suggestions, they were not asked to endorse the conclusions or recommendations nor did they see the final draft of the report before its release. The review of this report was overseen by R. Stephen Berry, University of Chicago. Appointed by the NRC, he was responsible for making certain that an independent examination of this report was carried out in accordance with institutional procedures and that all review comments were carefully considered. Responsibility for the final content of this report rests entirely with the authoring committee and the institution.

Contents

Summary

CONTEXT

In 1986, Congress established the U.S. Special Operations Command (SOCOM) as a new unified command to oversee training, doctrine, and equipping of all Special Operations Forces (SOF). Headquartered at MacDill Air Force Base, Florida, SOCOM is commanded by a four-star flag or general officer. SOCOM has four service component commands: U.S. Army Special Operations Command; Naval Special Warfare Command; Air Force Special Operations Command; and Marine Special Operations Command; and one subunified command, the Joint Special Operations Command. SOCOM's mission is twofold: 1) to plan, direct, and execute special operations in the conduct of the War on Terrorism in order to disrupt, defeat, and destroy terrorist networks that threaten the United States, its citizens, and interests worldwide and, 2) to organize, train, and equip SOF provided to geographic combatant commanders, American ambassadors, and their country teams. Today, SOF units include a total of approximately 55,000 active and reserve personnel across all the services, or about 2 percent of all U.S. active and reserve forces. SOF are specially organized, trained, and, equipped military forces that conduct operations to achieve military, political, economic, or informational objectives by generally unconventional means in hostile, denied, or politically sensitive areas.

The unique nature of SOF roles, missions, and tactics is always built on a single common denominator—the small unit. Within these small units, the capabilities and limitations rest on the individual skill sets of the team members and the interdependent functioning of the group. Whether functioning as an interdependent task force or semi-autonomously at a remote location, SOF units perform best with clearly defined objectives, prepared by intense training accompanied by detailed planning and rehearsals. SOF small units are uniquely adaptive and flexible, enabling them to quickly and successfully improvise and adjust to highly fluid situations and escalation of risk. At any given moment, SOF are likely to be engaged in some stage of the planning or execution of special operations in many countries around the world, spanning a wide range of environments and missions. SOF therefore must be capable of operating in environments ranging from tropical jungle to arctic, maritime to desert, subterranean to mountainous, and rural to urban. Within this vast range, additional factors may influence technical and operational requirements, including weather, topography, bathymetry, geology, flora, fauna, and human population density. All of these factors must be weighed in terms of the challenges they pose to supporting communications and operational security. In short, SOF must maintain the capability to operate globally, in any environment, against any threats that can be countered by its unique capabilities. Although SOF have broad and flexible capabilities to conduct missions that are also performed by conventional forces, SOF are distinguished from conventional forces in that they offer the best choice in situations requiring regional orientation, cultural and political sensitivity—e.g., military-to-military contacts—and for especially sensitive missions, including those likely to require covert and/or clandestine operations and operations involving access to denied areas.[1]

---

[1]William S. Cohen, Secretary of Defense, "Annual Report to Congress," 1998, Adapted from Chapter 4,

The focus of this report is primarily on the key core SOF task of special reconnaissance: on determining SOF-specific sensing and supporting communications (S&SC) needs and mapping them to existing and emerging technologies. The committee chose the special reconnaissance mission for this preliminary report because it traditionally employs a wide variety of sensing technologies over extended time periods to accomplish the mission. Also, the committee believes this mission is likely to have some of the most demanding requirements for time sensitivity, low probability of intercept/detection, and optimization of technology size, weight, and power, and therefore provides an upper bound for S&SC technology performance among all SOF missions.

SYNOPSIS OF PRELIMINARY OBSERVATIONS

This section contains a synopsis of the committee's preliminary observations that were deemed releasable to the public.

Observation 1: *While existing sensor systems can be tailored to support the SOF mission, the full spectrum of sensing capabilities remains to be exploited.*

Observation 2: *It is critical that accurate, very local weather information (both measurements and forecasts) be available to SOF personnel in the field, as well as to those who provide planning and support functions.*

Observation 3: *There is a lack of coordinated tasking (cross-cueing) between remote and close-in sensors.*

Observation 4: *The potential of sensor networks needs to be further exploited.*

Observation 5: *Unattended ground sensors and the man on the ground require improved communication techniques for infiltration and exfiltration of information.*

Observation 6: *Triple-canopy jungle sensor data exfiltration models must be better understood.*

Observation 7: *There is considerable need for communications models at all stages of a mission, for both exfiltration of data to command as well as local mission-focused communication.*

Observation 8: *Close coupling of sensor signal and information processing and supporting communications is required.*

Observation 9: *Multiple communications network architectures will be required to support SOF and their sensors.*

DISCUSSION

There was a notable contrast between the "mission-scale" technology emphasis presented to the committee in briefings by vendors and suppliers and the "man-scale" emphasis presented in vignettes by SOF operators. In particular, many of the concerns expressed by operators centered on size, weight, and power considerations that affect man-packability, and mission characteristics such as low probability of detection. These technology signature issues, while less of a concern against adversaries with low or moderate technological capabilities, are of significant concern in the case of missions with an adversary that is a technological peer of the United States or is equipped by one. Given the small size, independence, and missions of SOF teams at the "pointy end of the spear" and the possible inability to communicate for extended periods, it seems important to understand the technological capabilities that enhance a single warrior or small teams of warriors. Opportunities for advances related to sensing and supporting communications challenges in the jungle environment include new sensor capabilities, novel emplacement mechanisms, new sensor data exfiltration means, and innovative system architectures, including systems of relays.[2]

The committee is mindful that these technologies must function as part of a system that includes not only specific mission environments but also the particular, real-world needs of the SOF warrior. In addition, the committee anticipates that some of the restricted environments in which SOF operates—e.g., urban, jungle/triple canopy, mountainous, caves, and underground facilities—will impose more stringent demands on the types and performance of S&SC technologies than will more open environments—e.g., maritime, riverine, desert, arctic, and littoral/harbor. Further, it anticipates that technologies appropriate to the more restricted environments above will share many characteristics in common, as will those appropriate to the less-restricted environments. In conclusion, because a parallel NRC committee produced a separate in-depth report on radio frequency communication systems, the communications aspects of this effort were limited to communications in support of sensors.[3]

---

[2]An example of a new capability is the concept of the "electronic dog's nose." For additional information, please see M.J. la Grone, C.J. Cumming, M.E. Fisher, M.J. Fox, S. Jacob, D. Reust, M.G., Rockley, and E. Towers, "Detection of land mines by amplified fluorescence quenching of polymer films: a man-portable chemical sniffer for detection of ultratrace concentrations of explosives emanating from land mines," in *Proc. SPIE*, Vol. 4038, pp. 553-562, Dubey, Harvey, Broach, and Dugan, eds.: (Detection and Remediation Technologies for Mines and Minelike Targets).

[3]For additional information on the Committee on Universal Radio Frequency System for Special Operations Forces, please see the committee's public web site at www8.nationalacademies.org/cp/projectview.aspx?key=48926.

Appendixes

Appendix A

Biographical Sketches of Committee Members

Joanne Isham, Chair, is currently senior vice president, Washington Operations for L-1 Identity Solutions, and was formerly the chief operating officer of High Performance Technologies, Inc. Prior to assuming this role, Ms. Isham was the vice president and deputy general manager of Network Systems at BAE Systems. In addition to managing day-to-day operations, she was directly responsible for identifying strategic opportunities for new mission focus within the Intelligence Community, the Department of Defense, and the Department of Homeland Security. Before joining BAE Systems, Ms. Isham was a member of the Senior Intelligence Service and a career officer at the Central Intelligence Agency (CIA). From September 2001 until her retirement in 2006, she served as deputy director of the National Geospatial-Intelligence Agency (NGA). Immediately prior to that assignment, she served as deputy director for science and technology at the CIA. In that position she was the principal executive overseeing the CIA's scientific and technical program with particular responsibility for clandestine technical activities and agency research and development. Prior to that appointment, Ms. Isham served as CIA's associate deputy director for science and technology, the next most senior position in the directorate. Ms. Isham has held several other senior management positions in the CIA and other intelligence community organizations. She served as director of congressional affairs for the Director of Central Intelligence (DCI), overseeing the congressional and legislative interests of the DCI. She also served as the deputy director of the Resource Management Office of the Community Management Staff (CMS) and as CMS's director of program analysis. In these positions, she oversaw budget and resource issues spanning the entire intelligence community. Ms. Isham also spent a number of years on assignment to the National Reconnaissance Office, where she served as director of legislative affairs, a program monitor for national programs, and a program manager. She has received numerous awards throughout her career, including the National Intelligence Distinguished Service Medal, the Department of Defense Distinguished Civilian Service Award, the National Intelligence Medal of Achievement, the CIA Distinguished Intelligence Medal, the NGA Distinguished Intelligence Medal, and the DIA Director's Award. Ms. Isham is a member of the Senior Advisory Group for the Director of National Intelligence. She is a graduate of the University of Notre Dame.

Thomas Burns, Vice Chair, co-founded and serves as its CEO and chair. SET Corporation, an R&D company specializing in the development and commercialization of smart sensing technologies, prior to founding SET, he co-founded and served as COO of ObjectVideo Inc., a venture-backed leader in smart video solutions for commercial and military security applications. Dr. Burns joined ObjectVideo from DARPA, where he pioneered the development of model-based signal and image exploitation technologies, building on his experiences directing Computer Vision research as a U.S. Air Force Officer at AFRL. While assigned to AFRL, he led AFRL's premiere Automatic Target Recognition program, receiving AFRL's prestigious Peter R. Murray Program Manager of the Year award. Dr. Burns is co-inventor of patents on video and radar technology and has published numerous refereed papers in areas as diverse as electro-optics and wavelet mathematics. He holds a Ph.D. in electrical engineering from the Air Force Institute of Technology.

Kenneth Bowra is Oak Ridge National Laboratory's representative to the U.S. Joint Forces Command and serves as a senior mentor/concept developer at USJFCOM J9 for Joint Urban Operations and Homeland Security experimentation. He retired as a major general after 33 years of service in the U.S. Army. He was commissioned a regular army officer, in the infantry, after graduating from the Citadel. In his last assignment, he served as the assistant chief of staff, Operations, Allied Forces North Europe, NATO, where he led NATO's first out-of-Europe reconnaissance mission to Afghanistan. He has commanded Special Forces units at all levels including serving as a reconnaissance team leader in Vietnam with MACV-SOG and commander, 5th Special Forces Group (Airborne). General officer assignments include Commanding General, Special Operations Command South, U.S. Southern Command; Deputy Commanding General, U.S. Army Special Operations Command; Commanding General, U.S. Army Special Forces Command; Commanding General, U.S. Army John F. Kennedy Special Warfare Center and School; and Deputy Commander of NATO's Kosovo Force (KFOR). MG Bowra is the chairman of NDIA's Special Operations/Low Intensity Conflict Division. He is a member of the National Research Council's USSOCOM Standing Committee.

Regan Edens III is director of military and intelligence community support for Ares Systems Group, where he provides expertise in areas including research, development, testing, and engineering; special technical operations; special activities; and human intelligence, counterintelligence, and special operations. Mr. Edens attended Cornell University, College of Human Ecology. Cornell's Human Development program attempts to understand human behavior from a multidisciplinary approach combining the relevant aspects of biology, psychology, sociology, and governmental policy. In 2004, Mr. Edens was recruited to develop the first DOD Counterintelligence MASINT Collection Program for the Counter Intelligence Field Activity (CIFA). Mr. Edens was recruited to a temporary appointed government position with the Intelligence and Securities Command (INSCOM) for a new concept in rapid technology development and incubation in support of special operations, sensitive intelligence activities, and conventional forces. As deputy director for operations, Mr. Edens created training and operations concepts in support of rapid development, acquisition, and implementation programs of COTS, GOTS, and university research projects against high-value threats and intelligence, reconnaissance, and surveillance (ISR) collection requirements. Mr. Edens worked with engineers and government customers to develop the tactics, techniques, and procedures for mission integration. Mr. Edens led the first validation mission of the activity, for which he received high praise from the Coalition Land Forces Component Commander (CFLCC) in Afghanistan.

Chip Elliott is chief engineer at BBN Technologies and project director for GENI, a national-scale experimental facility being created by the National Science Foundation for "clean slate" research in global networking. He is an AAAS fellow and an IEEE fellow with over 85 patents issued and pending. Mr. Elliott led DARPA's design and build-out of the world's first quantum cryptography network—10 optical nodes across metro Boston providing highly secure key distribution nonstop through both telecom fibers and the atmosphere—as well as design and implementation of large-scale, mission-critical 'ad hoc' radio networks now used in nearly a dozen nations, including the United States, Canada, and the United Kingdom. For his leadership in quantum cryptography, he was given Frost & Sullivan's Award of Excellence in Technology (2005) and was named a World Technology Award Finalist (2004) and fellow. Mr. Elliott has served on many national panels, including the Defense Science Board; the Naval Studies Board (National Research Council); the Army Science Board; the Standing Committee on Research,

Development, and Acquisition Options for SOCOM; and the Technical Experts Panel for Quantum Crytography (DTO); he has held visiting faculty positions at Dartmouth College, Tunghai University in Taiwan, and the Indian Institute of Technology, Kanpur.

Carolyne Hart, Sandia National Laboratories, is a member of Sandia's Defense Systems and Assessments Program Director team and is the primary point of contact for Sandia's airborne and ground-based surveillance and reconnaissance activities. She is also a member of Sandia's satellite remote sensing and verification board of directors. In addition to her programmatic role, Dr. Hart is the line director of Sandia's Electronic Systems Center. As such, she is accountable for managing the laboratories' expertise in radio frequency electronics, communications and antenna subsystems, tagging, tracking, and locating, and synthetic aperture radar systems. Her organization is also responsible for the electronic subsystems in the nation's nuclear weapons stockpile and for the laboratories' precision pointing mechanisms, guidance, control, and motion compensation subsystems, flight computers, and real-time software for data acquisition and algorithms used in a variety of applications. Dr. Hart received a National Science Foundation scholarship for graduate studies at Oklahoma State University, where she received her Ph.D. in electrical engineering. As an adjunct professor at the University of New Mexico, she received the College of Engineering Outstanding Teacher Award on several occasions. She is a member of the IEEE.

Robert Henrick is the executive vice president, Customer Solutions, at Arbitron and is responsible for survey research methods and product management as well as enhancements of existing services and the delivery of new services serving the media industry. Dr. Henrick previously served as a program manager at the Johns Hopkins University Applied Physics Laboratory, leading the U.S. Navy Distributed Netted Sensors (DNS) ASW C4I Program. There, he led efforts in identifying and prototyping OTH and LOS communication solutions for DNS data exfiltration in military peacetime and wartime operations and acted as technical lead for ASW DNS systems studies and analysis of alternatives. Dr. Henrick managed efforts at APL to create high-gain, low-power ASW sensors for multiple applications using integrated MEMS and silicon RF ICs. From 1994 to 2003, he gained commercial experience in communications devices and services and Internet applications at AT&T, Lucent, Ogilvy and Mather Interactive, and several venture-funded start-ups. He holds 12 patents in the area of Internet and communications applications and services. From 1986 to 1994, Dr. Henrick co-founded the Bell Labs Washington, D.C., office and launched the DARPA Acoustic Warfare Program. Dr. Henrick holds a Ph.D. in applied mathematics from Rensselaer Polytechnic Institute.

Paul Hoff retired as director of advanced sensor and distributed fusion systems with BAE Systems, where he was responsible for organizing and managing cutting-edge technology teams in the development of integrated fusion and advanced unattended ground sensors and unattended air vehicles supporting military transformation. Dr. Hoff served as chief technology officer for company-wide initiatives in homeland security, with a focus on detection of weapons of mass destruction, and he directs the advanced research programs integrating industry, government laboratories, and major universities. Dr. Hoff has experience with the Army's Advanced Sensors Collaborative Technology Alliances programs focused on the development of integrated fusion and advanced unattended ground sensors and unattended air vehicles supporting military transformation.

Linda Katehi, a member of the National Academy of Engineering, is the provost and vice chancellor for academic affairs at the University of Illinois at Urbana-Champaign and professor of electrical and computer engineering. Prior to joining the University of Illinois, she served as the John A. Edwardson Dean of Engineering and Professor of Electrical and Computer

Engineering at Purdue University and as the associate dean for academic affairs and graduate education in the College of Engineering and professor of electrical engineering and computer science at the University of Michigan. As a faculty member, Professor Katehi has focused her research on the development and characterization of three-dimensional integration and packaging of integrated circuits, with particular emphasis on MEMS devices, high-Q evanescent mode filters, and the theoretical and experimental study of planar circuits for hybrid-monolithic and monolithic oscillator, amplifier, and mixer applications. She has pioneered the development of on-wafer integration techniques that led to low-cost, high-performance integrated circuits for radar, satellite, and wireless applications. Dr. Katehi has received numerous awards for her research contributions to the field of high-frequency circuit design. She is a member of the NRC Panel on Sensors and Electron Devices, a fellow of IEEE, and a member of Sigma Xi.

Robert McClatchey is president of McClatchey Associates and works as a consultant on applied meteorology and remote sensing. Dr. McClatchey consults on projects for private-sector firms on the subjects of atmospheric and surface radiation backgrounds, satellite remote sensing of the atmosphere and surface, and the radiometric effects caused by atmospheric trace species. He is an expert on the visible and infrared absorption, emission, and scattering properties of the atmosphere and the effects of weather on the modification of these properties. Dr. McClatchey currently teaches a graduate course on atmospheric radiation at MIT. Before his current work, Dr. McClatchey was the SES director of the Atmospheric Sciences Division of the Air Force Research Laboratory (AFRL), where he directed the work of over 100 scientists and engineers in a comprehensive R&D program. Dr. McClatchey's honors include SES Meritorious Executive Award, fellow of the American Meteorological Society, and fellow emeritus of the AFRL. He holds a B.S. in physics and an M.S. in meteorology from the Massachusetts Institute of Technology and a Ph.D. in meteorology from the University of California at Los Angeles. He has published over 40 technical papers.

Tamar Peli has over 25 years' experience in signal and image processing with application to DOD and the intelligence community (IC). Ms. Peli has developed technologies for a wide range of platforms and sensor modalities, including imaging and nonimaging radar from X band to low frequencies, IR/IRST/FLIR, video EO/IR, multispectral/hyperspectral, and MASINT. She is currently associate director, Special Operations Programs, Signal Processing & Exploitation for Draper Corporation. Ms. Peli is leading Draper's White Space Initiative in exploitation technologies. As the lead for the exploitation initiative, Ms. Peli is responsible for developing the technology roadmap, the business strategy for expanding Draper's role, and business opportunities within the IC/DOD community and its execution. Ms. Peli is an MSS National Symposium on Data Fusion committee member co-chairing the SCI session. She is also a member of the Panel on Digitization and Communications Science of the NRC's Army Research Laboratory Technical Assessment Committee. Ms. Peli holds an M.S. in electrical engineering from Technion—Israel Institute of Technology.

Ed Pentaleri is a communications payload systems engineering manager for Space Systems/Loral. He has spent more than 20 years leading and contributing to various aspects of technology development for important and challenging remote-sensing applications, including several DARPA-sponsored programs. His involvements have ranged from nonintrusive detection of contraband in luggage and cargo containers to classification of unexploded ordnance (UXO), detection of land mines, and the remote, automated detection and classification of chemical and radioactive wastes. More recently, he has led developments of large broad-band phased-array apertures for airborne COMINT/SIGINT and FOPEN GMTI applications. Although his background spans wide-ranging technologies and physical phenomena, the unifying focus has been on the development of practical solutions to important and demanding remote sensing and classification problems. He is a member of the IEEE Antennas and Propagation Society and the IEEE Nuclear and Plasma Sciences Society. Mr. Pentaleri holds an M.S. in physics.

W. David Sincoskie, a member of the National Academy of Engineering, is a professor of electrical and computer engineering at the University of Delaware. From 1996 to 2008, he was group senior vice president of the Network Systems Research Laboratory at Telcordia. The laboratory consists of over 100 researchers involved in many aspects of Internet and broadband networking. Major areas of activity in the laboratory are Internet network management, mobile and ad hoc Internet, wireless communications, and optical network management. Prior service includes the DARPA's Information Science and Technology Committee and the Internet Architecture Board. Dr. Sincoskie was executive director of the Computer Networking Research Department at Telcordia (formerly Bellcore) from 1990 through 1996. He managed a group working on the AURORA gigabit testbed, IPv6, IP over ATM, NSFNET, and broadband service control. Dr. Sincoskie is a fellow of the IEEE. He was inducted into the University of Delaware's Alumni Wall of Fame in 2006 and received the Distinguished Electrical Engineering Alumnus Award in 1994. He received the Bellcore President's Award in 1993. In 2003, he received the IEEE Fred W. Ellersick Prize paper award. He was an adjunct professor of computer and information science at the University of Pennsylvania from 1989 to 2008. Dr. Sincoskie currently serves on the National Research Council Board on Army Science and Technology (BAST) and was a member of the BAST Committee on Strategies for Network Science, Technology, and Experimentation.

Jonathan Smith is the Olga and Alberico Pompa Professor of Engineering and Applied Science at the University of Pennsylvania, to which he recently returned after almost 3 years at DARPA. He was elected IEEE fellow in the Class of 2001 for contributions to the technology of high-speed networking. He was previously at Bell Telephone Laboratories and Bellcore, which he joined at the AT&T divestiture. His current research interests range from programmable network infrastructures and cognitive radios to architectures for computer-augmented immune response. Dr. Smith serves on the Technical Advisory Group of the President's Council of Advisors on Science and Technology Network and Information Technology.

L. Danny Tromp is an associate technical director in MITRE's Electronic Systems and Technologies Technical Center, where he is responsible for overseeing the technical center's work in the areas of sensors, communications, networking, signal processing, and enabling technologies applied to electronic systems for command, control, intelligence, surveillance, and reconnaissance (C2ISR). He has over 28 years' experience in sensors and communication systems analysis and in applying signal and data processing to solve problems in intelligence, surveillance, and reconnaissance (ISR) applications. Dr. Tromp has contributed to a number of programs and has led several DARPA and MITRE Technology Program R&D projects. Dr. Tromp has led and contributed to several synthetic aperture radar, automatic target recognition, moving-target indication radar, tracking, and correlation and multi-INT fusion efforts. Other projects included development of foliage penetration (FOPEN) and counter camouflage concealment and deployment (CCC&D) technology to detect concealed ground targets, and advanced multisensor approaches to significantly enhance the ability to track, identify, and target surface moving targets. He has also worked in the areas of computer simulation, signal compression, spectral analysis, and electronic warfare. Dr Tromp is a senior member of the IEEE; he served on the Boston section of the IEEE committee. He was a member of the U.S. delegation invited by the National Academies and the oversight committee of the Global Dialogues on Emerging Science and Technology (GDEST) to speak at GDEST's first conference, Sensors and Sensor Systems: A U.S.-Japan Dialogue, which was held in Tsukuba, Japan. His current research interests include detection and estimation theory, spectrum analysis, statistical signal processing, and network-centric sensing. He has authored and published over 35 conference papers and technical reports. He holds a Ph.D. in electrical engineering from Northeastern University.

David Whelan, a member of the National Academy of Engineering, is vice president and deputy general manager of Boeing's Advanced Systems and chief scientist, Integrated Defense Systems. Prior to joining Boeing in 2001, Dr. Whelan was director of the Tactical Technology Office at the Defense Advanced Research Projects Agency (DARPA), where he led the development of enabling technologies, such as unmanned vehicles and space-based moving-target-indicator radar systems. Prior to his position with DARPA, Dr. Whelan held several positions of increasing responsibility with Hughes Aircraft. His high-technology development experience also includes roles as a research physicist for Lawrence Livermore National Laboratory, as well as one of four lead engineers assigned for the design and development of the B-2 Stealth Bomber Program at Northrop Grumman. Dr. Whelan is a member of the NRC USSOCOM Standing Committee.

Winston Wiley retired from the CIA as the associate director of homeland security. Previously he served as deputy director for intelligence. Earlier, he had held a number of key agency leadership positions, including deputy chief and, later, chief of the DCI Counterterrorist Center, and chief, Persian Gulf Division, where he headed the Intelligence Directorates Persian Gulf Task Force following Iraq's invasion of Kuwait. Mr. Wiley began his Agency career in the CIA's Career Trainee program. Following graduation, he was assigned to the Office of Current Intelligence and, later, the newly formed Office of Regional and Political Analysis (ORPA). After a year at Harvard University, Mr. Wiley was assigned to the Directorate of Intelligence planning staff and served as executive officer and research director in ORPA and, later, the Office of Near Eastern and South Asian Analysis. From 1983 to 1988, he served as deputy chief and chief of the International Security Issues Division, Office of Global Issues. He subsequently was selected for a rotation to the Inspector General's staff in 1988. During his career at the CIA, Mr. Wiley was awarded the Director's Award, the Distinguished Intelligence Medal, and the Donovan Award. Mr. Wiley received his M.P.A. from the John F. Kennedy School of Government at Harvard University. Prior to his professional agency service, he served 3 years in the former U.S. Army Security Agency.

Appendix B

Meetings and Participating Organizations

MEETING 1
JUNE 9-10, 2008
WASHINGTON, D.C.

Defense Intelligence Agency; Joint Personnel Recovery Agency; Naval Research Laboratory; Riverside Research Institute; Sandia National Laboratories; U.S. Special Operations Command

MEETING 2
JULY 17-18, 2008
WASHINGTON, D.C.

Defense Intelligence Agency; The Johns Hopkins University Applied Physics Laboratory; National Aeronautics and Space Administration; National Geospatial Intelligence Agency; Technology Services Corporation; U.S. Army Engineer Research and Development Center; U.S. Army Special Operations Command

MEETING 3
OCTOBER 9-10, 2008
WASHINGTON, D.C.

Air Force Research Laboratory; Army Research Laboratory; The Charles Stark Draper Laboratory; General Atomics; Harris Corporation; Johns Hopkins University Applied Physics Laboratory; Lockheed Martin; The MITRE Corporation Raytheon Company; Office of the Deputy Undersecretary of Defense (Advanced Systems and Concepts); U.S. Army

MEETING 4
NOVEMBER 17-18, 2008
WASHINGTON, D.C.

Naval Research Laboratory

MEETING 5
DECEMBER 17-18, 2008
WASHINGTON, D.C.

Writing meeting.

Appendix C

Suggested Reading

Bowden, Mark. Killing Pablo. 2001. Atlantic Monthly Press. ISBN: 0871137836.

DSB. The Future of the Global Positioning System. 2005. Washington, D.C.: Office of the Undersecretary of Defense for Acquisition, Logistics, and Technology.

DSB. Defense Biometrics Report. 2007. Washington, D.C.: Office of the Undersecretary of Defense for Acquisition, Logistics, and Technology.

DSB (Defense Science Board). 2008. Joint Defense Science Board and Intelligence Science Board Task Force on Integrating Sensor-Collected Intelligence. Washington, D.C.: Office of the Undersecretary of Defense for Acquisition, Logistics, and Technology.

NRC. 2004. Army Science and Technology for Homeland Defense: C4ISR. Washington, D.C.: The National Academies Press. Available from http://www.nap.edu/catalog.php?record_id=11053.

NRC. 2004. Summary of the Sensing and Positioning Technology Workshop of the Committee on Nanotechnology for the Intelligence Community: Interim Report. Washington, D.C.: The National Academies Press. Available from http://www.nap.edu/catalog.php?record_id=11032.

NRC. 2005. C4ISR for Future Naval Strike Groups. Washington, D.C.: The National Academies Press.

NRC. 2005. The Navy's Needs in Space for Providing Future Capabilities. Washington, D.C.: The National Academies Press.

NRC. 2007. Distributed Remote Sensing for Naval Undersea Warfare: Abbreviated Version. Washington, D.C.: The National Academies Press. Available from http://books.nap.edu/catalog.php?record_id=11927.

NRC (National Research Council). 2009 (in press). Toward a Universal Radio Frequency System for Special Operations Forces: Abbreviated Report. Washington, D.C.: The National Academies Press.